STOTHERT & PITT

THE RISE AND FALL OF A BATH COMPANY

STOTHERT & PITT

THE RISE AND FALL OF A BATH COMPANY

John Payne

Millstream Books

*The book is dedicated to our first grandchild, Megan,
born 6 March 2007,
who has yet to discover the excitement of cranes.*

Dinnertime in the tool-room. A game of shove ha'penny with, from left to right, Fred Bradfield, Bob Potts and Glyn Phillips. (Brian Higgs collection)

First published in 2007 by Millstream Books, 18 The Tyning, Bath BA2 6AL

Text set in 11 point Gill Sans Light, headings in 21 point Copperplate Regular

Printed in Great Britain by Short Run Press Ltd, Exeter

© John Payne 2007

ISBN 978 0 948975 79 0

British Library Cataloguing-in-Publication Data: a catalogue record for this book is available from the British Library

All rights reserved. No part of this publication may be reproduced, stored in a retrieval system, or transmitted in any form or by any means electronic, mechanical, photocopying, recording or otherwise, without the prior permission of Millstream Books.

Foreword

The story of Stothert & Pitt is a perfect example of the capacity of history to confound expectations. Bath is a city internationally renowned for its eighteenth-century architecture and its air of sophisticated elegance. The very fact that the city has an industrial heritage is surprise enough to most visitors (and some residents) but the existence of a large engineering business, which flourished for over 200 years and which, at its height, employed thousands, seems to many, encouraged to think of Bath simply as a holiday destination, hard to believe.

Many of the physical remains of Stothert & Pitt's activities in Bath have disappeared and those that do remain, a few buildings and a bridge, are under threat of development or neglected. The history of the firm is overlooked. This is unfortunate as from the 1780s until the 1980s Stothert & Pitt played a major part in the working life of the city and made famous a range of equipment for the civil engineering, construction and transport industries which had a global market.

The surviving records of Stothert & Pitt have been preserved at the Museum of Bath at Work and include over 40,000 photographs and drawings, recorded interviews with former employees and products. There is, in addition, a huge collection of catalogues and literature. Sadly a combination of river flooding and bombing during the Second World War destroyed much more.

Although the museum holds two scale models of cranes, the sheer size of the equipment produced by the company makes a display of its larger manufactures impossible. There is an irony in this: Stothert & Pitt products, be they dockside cranes, road-rollers or cement mixers, are scattered across the world, many still in use and well maintained, while the city where this equipment was produced has nothing substantial on show. Nevertheless the full story of the firm is told at the Museum of Bath at Work.

In 1980 the director of the Science Museum stated that Stothert & Pitt was Bath's most important contribution to world history. For those happy to see the industrial heritage of Bath forgotten or written out of the city's history, this should be well borne in mind.

Stuart Burroughs
Director, Museum of Bath at Work

Acknowledgements

Stuart Burroughs made available to me the considerable archive of Stothert & Pitt material in the Museum of Bath at Work. Don Browning made copies of many of the black-and-white images used in this book and Robert Coles scanned others. I would like to thank them all. Photographs in this book without attribution come from this collection.

I also wish to thank Sylvia Marks for material about the Maxwell Pensioners' Action group, and other Bath people with whom I discussed life at Stothert & Pitt. Gerry Matthews of M&B Engineering in Bellott's Road allowed me to photograph the one remaining Stothert & Pitt site still involved in engineering production. David Stephenson at Bath & North East Somerset Council kept me informed on the various options for the future of the Victoria and Newark Works. Beryl Leigh was instrumental in locating a copy of the 1885 Stothert & Pitt catalogue in the British Library and we are grateful for permission to reproduce some plates from this book.

Ian Guy told me about the Stothert & Pitt Rugby Club, and Dave Barnes and Albert Davenport of the Stothert & Pitt Bowls and Tennis Club told me about the continuing use of Newton Fields for sporting activities.

Brian Higgs talked to me about life in the tool-room and *The Spectres* (below) and provided photos, as did Dave Barnes, Ted Nixey, John Skinner and John Woodward.

Matthew Zuckerman of the *Bath Chronicle* arranged for the publication of some of the photos in this book. This led to many people contacting me to identify old friends, relations and workmates. They include David Ball, Dave Barnes, Alan Bateman, June Burford, Colin Carless, Maurice Cottell, Myra Douglas, Peter and Sandra Edgell, Betty Flint, Ian Guy, Brian Higgs, Keith Letts, Steve Lewis, Dave Lord, Gary Luton, Ted Nixey, Phil Richards, John and Ruth Skinner, Mervyn Toogood, John Willcox, Don Withers, John Woodward and Brian Wynes. I apologise in advance if any of the identifications or spellings are incorrect.

Finally, I would like to thank Tim Graham at Millstream Books for supporting this project and for his splendid editing of text and pictures.

John Payne
Frome, August 2007

The Spectres, 1966 line-up. (from left to right) Bob Pierce (lead guitar, Machine Shop), John Derrick (bass guitar), Brian Mould (vocalist), Chris Jeffreys (rhythm guitar and sax), Brian Higgs (drums, Tool Room), Brian Neathey (rhythm guitar). The band met up as boys at Oldfield Park Baptist Church Youth Club. Two of them worked at Stothert's. 'Mouldy', their lead singer, later died of a heart attack, and a CD by the group raised £5,000 for the British Heart Foundation. The Spectres still play at 60s evenings and private parties in the Bath area. (Brian Higgs collection)

CRANEMAKERS TO THE WORLD

On Raratonga, the largest of the Cook Islands with a population of 10,000, three hours by plane north of New Zealand, is a watering-hole called Trader Jack's. Mine host is loud and encouraging and forthright, and there on the wall is a metal plaque marked 'Stothert & Pitt Bath England 1918'. It had come from an old dockside crane and belonged to one of the locals. Such was the reach of Stothert & Pitt, crane-makers to the British Empire. Unfortunately Trader Jack's blew down in a cyclone before I could check this fact. A normal event in that part of the globe. It will be rebuilt, and maybe the Stothert & Pitt plaque will reappear. Stothert & Pitt were truly 'cranemakers to the world', in the appropriate sub-title which Ken Andrews and Stuart Burroughs gave to their 2003 book about the firm.

What will not reappear is the great company which at its peak employed 2,500 workers in Bath, and has its roots as firmly in eighteenth-century Bath as any of the Georgian buildings which justify Bath's status as a World Heritage Site. By 1985 Stothert's was in trouble, as manufacturing industry collapsed all across Britain. Lame ducks were being slaughtered left right and centre, sacred cows were discarded without further thought, and pigs might fly for all that the government cared. The TUC organised 'The People's March for Jobs' but unemployment went on rising, peaking at over three million. It had always been assumed in postwar Britain that such a figure was unimaginable, and that if it did happen, revolution would follow. Yet in 1985 the miners returned to work after the best part of a year on strike, and Mrs Thatcher had beaten the most militant (if not the best organised and led) section of the British workforce. At Stothert's a new Chairman was appointed – George King. The workforce was down to under 1,000, another 160 jobs were about to go. Losses were piling up, millions of pounds' worth of them.

In October 1986, ex-Labour MP, millionaire, playboy and tycoon Robert Maxwell, through his Hollis Group of engineering companies, took control of Stothert & Pitt. Maxwell's core interests were in publishing, but Maxwell represented too the strike force, the cutting edge of the new economy which was about to emerge from the ruins of the old. Its critics called the activity asset-stripping (or worse). A very rich man or company would buy up an ailing business, hasten its demise, and then make a profit by selling off any viable assets, whether an individual plant, land or whatever. Stothert & Pitt had land, a great swathe of it along the River Avon in Bath, as we shall see. It also had a healthy pension fund, of which more later. By May 1987 George King was suing for wrongful dismissal – he eventually got £75,000 compensation. Shareholders were told the company was profitable. By August they were declaring losses of £3.84 million. In October 120 new jobs were announced, balanced by the closure of the Bitton works near Bristol. But the firm complained that most of those applying for the new jobs were unqualified. In July 1988 Maxwell sold Stothert & Pitt and the rest of the Hollis Group for £115 million in a management buy-out. A nice return on his 1985 purchase price of £4 million.

On 11 January 1989, the closure of Stothert & Pitt was announced. As many had suspected, it had simply not been possible for the new management team to meet the interest payments on the loan secured to buy the firm. *The Bath Star* reported 'furious allegations of asset-stripping and unfair behaviour'. It was just not cricket, as employee David Lord pointed out:

> *I am absolutely shattered. Stothert & Pitt has been my whole life. I've played in the cricket team and helped in the social club and now they have just come in and wiped it all out with a single strike.*

He received £4,500 redundancy pay after 38 years, whereas George King had received £75,000 for a couple of years' work. The council condemned the closure, but what else could they do? Another local paper, *The Bath Chronicle*, published pictures of angry workers outside the Guildhall with the illuminated Bath Abbey in the background – two contrasting images of the city in one press photo. Office worker John Hanham, 63 years old, collapsed and died a few hours after receiving his redundancy notice. His son-in-law said:

> I cannot believe that someone just gave him an envelope with his name written on the side, just like someone who'd been there two weeks.

One ex-employee, 77-year-old Albert Nurse, pointed to the personal significance of working for Stothert's:

> To be an engineer made one feel more of a man. Good work-mates made life enjoyable in spite of the noise, smoke and dust.

He was the third generation of his family to work at Stothert & Pitt. His grandfather had walked from Bristol in 1886 to get a job there. The swift and sudden end of Stothert & Pitt was a blow not just to people's pockets, but to their sense of self-esteem. The men felt it especially badly. To be a Stothert's man or woman was a way of belonging in Bath, your credentials in the community.

(right) Dave Barnes and Malcolm Kimmidge lead a trade union delegation to the Guildhall in January 1989. The 'two Robbies' were Robert Maxwell and Colin Robinson (who had lead the management buy-out). On the right of the photo is Adie Lyons. The four main unions were the Boilermakers, the Amalgamated Union of Engineering Workers (AUEW), the Transport and General (T&G) and the Electrical Trade Union (ETU). Each union had a Works Convenor, and they jointly formed the Works Committee. Prior to the purchase of the firm by Robert Maxwell, industrial relations had been good. As Dave Barnes, the Boilermakers' Convenor, put it: 'They used to listen to us and take on some of our suggestions.' The style of management changed abruptly when Maxwell took over. (Bath Chronicle)

Within two years, Maxwell was dead. It soon emerged that the Mirror Group's debts vastly outweighed its assets, and £440 million was missing from its pension funds. No-one has ever been prosecuted successfully for these crimes, although there must have been people who knew what was going on. As economic power in Britain passed from the world of industry and commerce to that of finance and banking, many City institutions were implicated in this massive fraud. The government's response was to continue deregulating the City: Maxwell was just one of a number of high-profile financial scandals in 1990s Britain. Stothert & Pitt pensioners suffered like others in the Mirror Group during the long campaign to get restitution of their rights. The basic claim was that because the 1986 Financial Services Act had removed the supervision of pension funds from the Department of Trade and Industry, it was up to the government to take responsibility for the fiasco. Over three years, the Maxwell Pensioners' Action Group lobbied MPs and government ministers. And won, after three long years of campaigning.

Yet the only acknowledgement of guilt was a £276 million payment by a raft of City institutions and what was left of the Mirror Group to the various pension funds, topped up by £100 million from government. Another version of this would be £100 million from me and you – the taxpayers of Britain. Another way to express it would be £2 per head from every man, woman and child in Britain to pay for Maxwell's crimes. In retrospect, Maxwell was not an isolated criminal, and the robbery of the pension funds of workers was a direct result of the government's over-dependence on the free market, and failure to provide adequate safeguards for industry in general and financial services in particular. Like so much in modern Britain, the free market created wealth for the few and misery for the many. That wealth would no longer be dependent on the manufacturing activities of firms like Stothert & Pitt but on the soulless world of banking, insurance and the Stock Exchange.

Imagining the City

Bath has an image problem. Or rather, it has an image which reflects certain aspects of the life and history of the city, but consistently ignores and downplays others. Into the first category we might place the Roman Baths, the spa waters, the legend of King Bladud, John Wood (father and son), Beau Nash, Georgian buildings and, coming closer to our own time, the Bath Festival, Bath University on its infrequent descents from the heights of Claverton Down, and the brand new spa centre with its pseudo-Latin name of Thermae Bath Spa. In short, the Bath that is known by visitors and tourists. What we might include in the second category is Bath as a place to live and work, the radical politics of the Parliamentarians in the seventeenth century, parliamentary reform and Chartism in the nineteenth century, the turbulent life of suburbs such as Walcot and Twerton, botched development decisions, pubs and clubs and factories and offices. In short, the Bath of residents and workers. It is a persistent dichotomy in views of the city. The poet Sylvia Townsend Warner recognised this half a century ago in her 1949 book *Somerset*: she emphasises medieval Bath as 'a place where people lived, rather than where people visited', personified by Chaucer's industrious Wife of Bath.

When upper-class England adopted Bath as its playground of preference in the eighteenth century, it was with some sense at least of Bath's history. At first, those promoting the city as a place of pleasure and healing emphasised the heroic British past, the legend of Bladud and the miraculous healing qualities of the warm springs gushing from the earth. This in turn gave way to John Wood's attempts to synchronise his building plans for Bath with the evidence of earlier civilisations in the area, such as the giant circles of standing stones at Avebury, Stanton Drew and Stonehenge. His architecture makes conscious reference to this heritage, which he also traced to buildings such as the Coliseum in Rome and the Temple in Jerusalem. As the century marches forward, and Bath becomes more respectable, Bladud fades back into the pre-Roman murk and it is the orderly, classical heritage of Ancient Rome that stands supreme. Beau Nash, in this reading of the recent past, is remembered as the man who set the rules by which polite society was to conduct itself in Bath, rather than the enterprising young man who made a fortune at the gaming tables.

What was leisure for wealthy residents and visitors was, of course, work for many in the city. The village of Weston established itself as the laundry-house of the city, with a fearsome reputation (at least during the eighteenth and nineteenth centuries) for strong, independent women who regarded the Sabbath as an occasion for the worship of strong drink and sexual licence rather than as a day of prayer and religious observance. John Wroughton, in his beautifully crafted essay on 'Bath and its Workers' demonstrated the extent to which the supply of personal services continued to be a major source of work in Bath into the nineteenth century. At the 1841 census, this group of domestic servants, gardeners, cooks and so on, constituted one third of the total working population of the inner parishes of Bath. One in nine worked in the fashion industry, and a similar proportion in the numerous handicrafts that produced luxury goods. By the time the census enumerators had worked through the various professions that made up the population of this city of leisure, only one in nine could be classified as industrial workers, including those working in the city's breweries, in the cloth mills at Twerton, and on the construction of the Great Western Railway.

The radicals of Bath, many of them craftsmen, were active in the agitation that preceded the Reform Act of 1832. Both Chartism and the Anti-Corn Law movement had active supporters in Bath through the 1830s and 1840s. From 1832-37 and again from 1841-47 Bath elected the Radical John Arthur Roebuck as one of its two MPs, much to the disgust of the Tory *Bath Chronicle*. In 1832, George Stothert, ironfounder, was one of 50 Bathonians who signed a letter to *The Times* in support of Roebuck. The historian R.S. Neale described the period from 1812-47 in Bath as 'a radical utopia'. But there was no large social base of industrial workers to support such a vision of utopia, and the wealthy of the city made sure that there was sufficient charitable provision to mop up the worst of social and economic distress. As a Chartist poster of 1841 put it: 'The thief throws the dog a bone to keep it from barking: you are to be quiet when good people are kind and generous.' From the mid-nineteenth century, Bath consistently returned Conservative Members of Parliament, and it was not until the revival of the Liberal Party, now the Liberal Democrats, in the final years of the twentieth century, that this changed. Trade unions were active in the city, not least at Stothert & Pitt, but were generally unable to command support for militant industrial action.

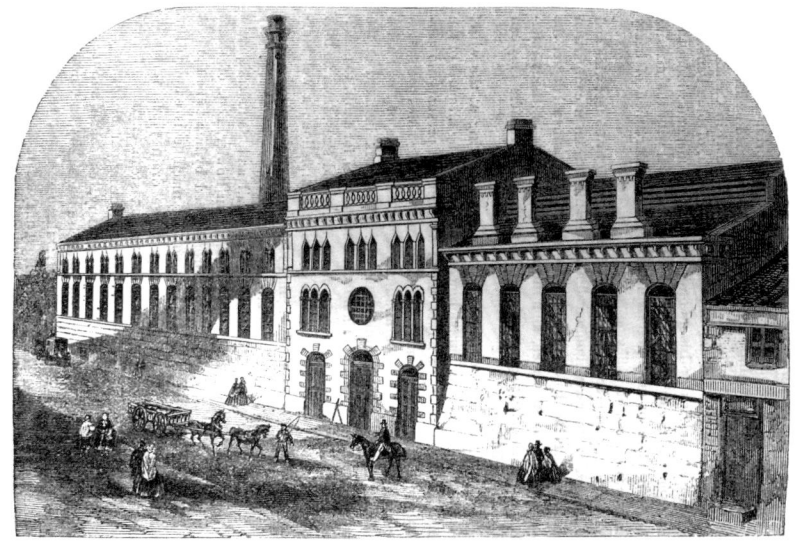

(above) An engraving of 'The New Premises of Messrs Stothert and Pitt' from The Official Illustrated Guide to the Great Western Railway *by George Measom, published in 1861. It is interesting to compare this with the engraving overleaf.*

(overleaf) The frontispiece of the 1885 Stothert & Pitt catalogue. Described as an 'Illustrated and descriptive price-book of machinery and ironwork', compiled by Cornelius Cornes and T. Calvert, and beautifully illustrated throughout with fine line engravings, the catalogue sold for 10/6 (half-a-guinea in Victorian terms, or just over 50p in modern terms). In just over 300 pages it gives a very complete account of the products which Victorian Stothert & Pitt made and sold. The frontispiece shows a fine view of the Newark Works in the Lower Bristol Road. (British Library, London)

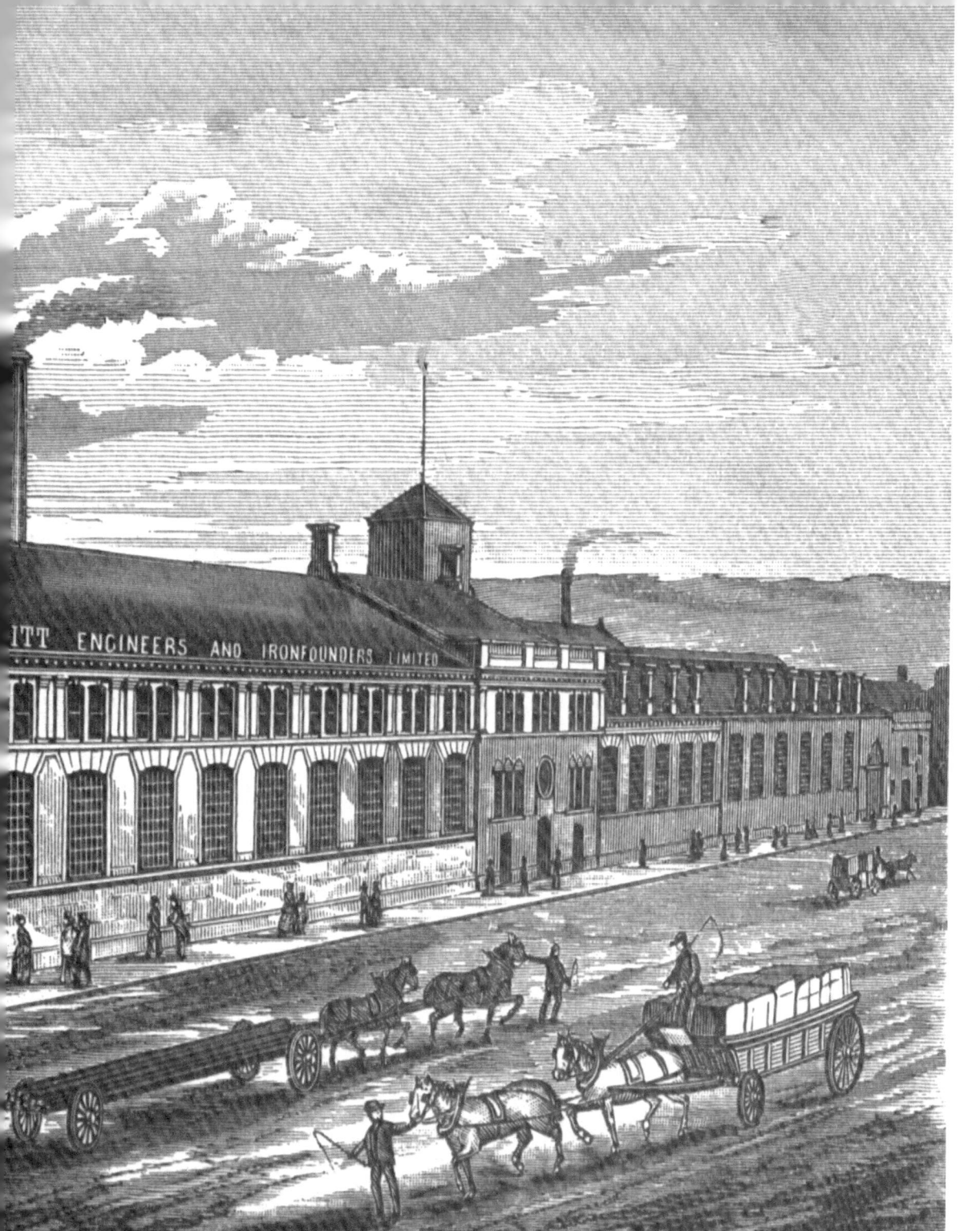

A Short Walk South of the River

The remains of Stothert & Pitt stretch for nearly two miles along the Lower Bristol Road out of Bath. For older residents of the city, and especially for those who worked at Stothert & Pitt, there are many memories to be recollected.

We begin our short walk just across the Churchill Bridge, in the shadow of the Great Western Railway viaduct and Brunel's mock Gothic gateway which allowed south-bound traffic to continue to flow across the River Avon as in centuries past. The precipitous, wooded heights of Beechen Cliff brood down. Looking back across the river to the Broad Quay, only a couple of narrow-boats that have presumably worked down through Widcombe Locks and the Kennet and Avon Canal act as a reminder that this was once the trading heart of Bath. The transport of goods up and down the river to Bristol and via the canal to Reading and London gave rise to the development of important industries here in the nineteenth century – a brass foundry, iron works, brewery, malthouses, a shopfitting firm, slaughterhouse and dye-works.

The industrial buildings in turn butted on to the teaming slums of Avon Street and Milk Street, originally built as part of 'Georgian' Bath, but too close to the river, too damp, too unhealthy to survive in any other form. It was to this part of town that my great-grandfather moved in the mid-nineteenth century, breaking one small link in the chain of generations of farm labourers who had worked on the Waldegrave estates at Chewton Mendip. This part of Bath's heritage has all gone, replaced by a confused mixture of social housing, a brutalist College of Further Education, a car park – the sum of commercial pressures and lack of foresight on the part of the City Fathers in the second half of the twentieth century.

Moving west along the Lower Bristol Road, past the red brick of the Bayer corsetry factory, we come to the first physical vestiges of Stothert & Pitt. The Stotherts moved to Bath from Shropshire in the latter years of the eighteenth century. George Stothert was a friend of William Smith, the geologist, who based his theories of rocks and their evolution on such practical tasks as sinking mine-shafts in the Somerset coalfield and the construction of the Somersetshire Coal Canal just south of Bath. The Stotherts used their Shropshire contacts to become agents for the Coalbrookdale Iron Works, who supplied the two elegant little wrought-iron footbridges across the Kennet and Avon Canal in Sydney Gardens and at the top lock at Widcombe. At this stage, the business was mainly retailing – fire grates, oven doors, and other domestic items. But from 1815 retailing was supplemented by manufacturing, an early commission being the House of Correction (prison) at Shepton Mallet. Robert Pitt joined the firm as an apprentice in 1834, became a partner in 1844 and Stothert & Pitt was born.

The Newark Works, designed by local architect Thomas Fuller, is the largest and grandest of the empty industrial buildings between the Lower Bristol Road and the river. It was built in 1857, its Bath stone front speaking of a firm that had every intention of becoming as important a part of the Bath landscape as the crescents and terraces designed by John Wood (father and son). The late 1960s Oak Street office block (opposite the corsetry factory) still looks smugly new today, and is still in use as offices, though not those of Stothert & Pitt. Looking at this rather elegant, confident office block, on stilts to provide car-parking and to protect against possible flooding, it is difficult to imagine that 20 years later the firm would have collapsed so completely.

The Newark Works stand proud along the Lower Bristol Road, nestling next to the Bayer corsetry factory and below the grand terraces and crescents of Georgian Bath. (Andrew Swift collection)

Opposite the Newark Works are the remnants of the busy goods yard of the Great Western Railway, which connected Bath to the wider world more surely and more swiftly than rivers and canals and turnpikes had been able to do. Past the *Green Park Tavern*, some of the cramped little terraces of working-class housing still exist. Other houses were taken over as office accommodation by Stothert & Pitt and have been subsequently replaced by substantial office blocks. Opposite the cemetery used to be the goods yards of Bath's other railway, the Somerset and Dorset. Now there is a one-way road scheme, Sainsbury's and its DIY companion, Homebase. The Sainsbury's garage occupies the site of the proudly modern (in the late 1950s) Stothert & Pitt Canteen where workers from the various factories and offices within walking distance could get a hot, cheap dinner.

(above) Thomas Fuller's somewhat eclectic design for the central part of the Newark Works. (Author's collection)

(below) The rear of the Newark Works in the floods of 1925. Flooding continued to be a problem for much of the life of Stothert & Pitt with serious floods occurring again in 1960. The curious design of the Oak Street office block on stilts was one attempt to solve the problem. Flooding is one reason why many of the records of the firm have been lost over the years.

(below) Flood scene at the new canteen on the Lower Bristol Road, 5 December 1960. The canteen was a recent addition to social welfare facilities at Stothert & Pitt. It is now the site of Sainsbury's petrol filling station. Sufficient of Georgian Bath is visible in the background to form an interesting contrast with the mid-twentieth-century 'fit-for-purpose' canteen building. Its destruction is regrettable, given that Bath has so few examples of good, functional twentieth-century buildings.

Aerial view of the Victoria Works (the area now known as the Western Riverside), looking across the river to the Lower Common and the Royal Victoria Park. The Somerset and Dorset Joint Railway is in the foreground. The surviving terraces of small stone houses suggest a date in the 1920s or 1930s.

Further west, a continuous terrace of houses blocks off the view to the north. Behind this terrace, and beyond what used to be the Somerset and Dorset Railway, the Victoria Works of Stothert & Pitt were laid out, with their own railway sidings, as the firm became one of the largest manufacturers of cranes in the country. Wherever the British Empire spread, Stothert & Pitt spread too. There was housing here, but little of it survived the remorseless

The new boiler shop at the Victoria Works, c.1924. The Victoria Works gradually expanded before, during and after the Second World War to fill the land between Victoria Bridge Road and Midland Road, with direct access to the Somerset and Dorset Railway.

expansion of Stothert & Pitt. An early site of Bath City Football Club, behind the *Belvoir Castle* pub, disappeared into the works. This area is now known euphemistically as the Bath Western Riverside, including the old Bath Gasworks with its giant landmark gasometers. Exactly what mix of industry, retail, leisure and housing this site will include, almost 20 years after the demise of Stothert & Pitt, is now becoming clearer. Housing and retail, yes, but little else besides. Given the long history of controversial 'grand projects' in Bath – the long-delayed and absurdly expensive Bath Spa project and the Combe Down quarries are just the most recent – it is hard to feel optimistic.

Eventually, no doubt, Stothert & Pitt will disappear from the face of Bath, just like the great curving viaduct that carried the Somerset and Dorset Railway up over the Lower Bristol Road and Windsor Bridge Road, and within a few feet of the first floor windows of the *Royal Oak* pub. The pub survives and is recently restored. The viaduct is gone. Beyond this, Bellott's Road turns uphill to the left. Stothert & Pitt had a small factory on a cramped site between East Twerton cemetery and the Great Western Railway, and the first of my many holiday jobs with Stothert & Pitt as a student was in the stores here. The factory is still in use, and by an engineering firm too – M&B Engineering. If the Somerset and Dorset viaduct across the Lower Bristol Road had to go, it was clearly easier to simply leave the one at the top of Bellott's Road, which crossed the Great Western Railway at this point. Despite its bad press – S&D came to stand for Slow and Dirty – its engineering structures were built to last! The track bed between here and the Devonshire tunnel is now the quaintly named Linear Park, full of birds and flowers and dog-walkers and splendid views of the Georgian city across the valley.

Further west again, there are more small stone terraces (Vernon, Argyll, Albert), the last of four pubs – the *Golden Fleece*, its name redolent of the Twerton weaving industry – and more new office buildings. Beyond the *Golden Fleece*, the Lower Bristol Road runs next to the smoke-blackened Great Western Railway viaduct, which sliced Twerton effectively in half. Some of the displaced residents in the nineteenth century were rehoused in damp hovels built into the railway arches. Carr's woollen mills, a major Victorian employer, have gone. But the brewery building remains, now skilfully renovated as the headquarters of Somer Housing, currently the major social housing landlord in the unitary authority of Bath & North East Somerset. Here is McDonald's too. A road sign reads 'River Place', but the tiny row of cottages of that name squeezed between the mill and the river was demolished years ago, as were the mills. Here at the end of the nineteenth century, my Payne grandparents raised 10 children (some say 11 – our family history is like that) between emerging from the slums of Milk Street and ascending to the respectability of Oldfield Park. Stothert & Pitt's own contribution to Twerton remains standing, a rather gaunt 1960s block built to house the Avalon Training School for apprentices. In its own very modest way, it is a tribute to an era which had rather more faith in the future than we allow ourselves today.

Beyond the apprentice school, a bailey bridge gives access to what was surely the most romantic of the Stothert & Pitt sites – Weston Island, between the Twerton weir and the canal cut that led barges through a lock and past the Dolphin Inn. On this island, the hard-working employees of the stores department sometimes did what we called 'physical checks'. Perhaps the large ledgers we kept by hand were unclear about how many crane jibs of a particular size were lying around the island. We would pick our way carefully through the nettles and elder bushes and rosebay willowherb to find which pieces of rusting iron bore the correct part code for which we were searching. It should not be possible to lose a crane jib – but on Weston Island anything was possible. Until they built a road around it that put paid once and for all to the wildlife and the excitement.

There is a little more of Stothert's, but the reader may need a car or bike to reach it. Newton Field, just before the point at which the Lower and Upper Bristol Roads finally unite, was the Stothert & Pitt sports ground, with facilities for football and rugby in winter and cricket, tennis and bowls in summer. I used to go out on my bike with Tony Marks to spot trains (the Great Western Railway line borders the field) and play tennis (a game his family played well, but which I never mastered). I was good at sums, liked sports, but was not good at them. That meant that at Newbridge Junior School I shared the cricket scoring duties. I enjoyed it. I could even work out end-of-season averages. From there I graduated to scoring for Bath Junior Schools – at Newton Field. The Bristol team had to bring a teacher to score for them. Later on, Tony and I would occasionally turn up on a Saturday afternoon in summer to change the manual score-board on the pavilion when the Stothert & Pitt cricket team were playing at home.

For the really brave, the search for Stothert & Pitt might be continued on along the Bristol Road nearly to Saltford. There in the meadows between the road and the River Avon is the Stothert & Pitt Rugby Club, a splendid tribute to the resilience and foresight of its members. They were always called The Cranes, and the name has stuck. A crane appears on their flag, and the link between the long-legged bird and its mechanical counterpart is at once apparent.

A SELECTION OF STOTHERT & PITT CRANES AROUND THE WORLD

(above) 4-ton electric gantry cranes at Riachuelo, Buenos Aires, Argentina in 1910. British trade before the First World War was not limited to the British Empire, and there were strong commercial ties with Argentina. No doubt the cranes were used to load Argentinian beef for the British market, among other goods. Although the first Stothert & Pitt electric crane was supplied to Southampton in 1893, the photograph opposite, above, suggests a continuing demand for steam-driven cranes.

(right) A screw-luffing crane on test at the Newark Works. This is a rare view of the yard at Newark, looking across the river Avon to the Avon Street flats. The truck suggests a date in the 1920s or 1930s.

(above) Two Stothert & Pitt veterans photographed at Colombo Harbour, Ceylon (Sri Lanka) in 1949. A 33-ton steam-driven block-setting Titan crane, built in 1898, and a 35-ton crane, also steam-driven, built in 1907.

(left) A container crane in course of erection at Rotterdam in 1967. As the British Empire began to crumble, markets in Europe and beyond became increasingly important. The 'export drive' was a national priority and Stothert & Pitt played a leading role in this.

Stothert & Pitt in the Second World War

Two pictures of bomb damage in the Structural Shop, Victoria Works in 1942. Since Stothert & Pitt was engaged in war-work, the factory might be regarded as a legitimate target. The same would not apply to the many public buildings and houses in Bath destroyed by Second World War bombs.

(above) A seaplane crane aboard HMS Ark Royal, off Plymouth Hoe in 1940, hoisting a reconnaissance plane from the water. Military customers were always important to Stothert & Pitt.

(left) An exhibition at the Victoria Art Gallery, Bath in 1945. Stothert & Pitt had made a great contribution to the war effort. With its cranes, concrete mixers, vibrating rollers and pumps, it was now set to make its contribution to the peace. The country was bankrupt and Stothert's, like other companies, encouraged its employees to save on a regular basis.

Stothert & Pitt: Biography and History

I am 19 years old. It is the spring or even the late summer of 1964. Or maybe of 1965. It is a fragment of memory. I am working in the stores at Weston Island. There are three or four of us. One of them I know. Like me, he is a regular at the *Dolphin*. He talks constantly of the Buffs, the Royal Antediluvian Order of Buffaloes. It is a Masonic organisation, and I thought masons were secretive. This must be the most talkative mason ever. The boss, the supervisor, the foreman (I think that was the term) is from the other side of Bath, out Larkhall way, unknown territory for a young man from Lower Weston. On my first morning, I finish the allotted task five minutes short of the allocated tea-break, and get my newspaper out. Immediately he is on me. The work must fill the time available (who said that?). That is clearly the most important rule here. As I relax into the job, get to know the men on the shop-floor, begin to explore our island and its treasures of wild flowers and rusting metal, so he, the foreman, relaxes too. He is something to do with the rugby club, The Cranes, and this takes time. No wonder his office needs a little extra help (me).

Monday is a busy day. There are various views of the team's performance the previous Saturday to canvass; there is a casualty list to review of injuries old and new; there are travel plans to make for the coming weeks' away fixtures. Friday is equally busy: again a casualty list to review, final team selection to be settled, travel arrangements to be confirmed, last minute disasters to deal with – a death in the family, a sick child. It is a busy place, our little stores office perched high above the shop floor. We may not be running the firm, but with the foreman's active involvement, and the tacit agreement of the rest of us ('Sorry, he can't talk at the moment, he's in the middle of an urgent call') that is what it feels like. We discuss the difference between 'nicking' and 'stealing'. Nicking is taking a few washers to boost up a sagging metal gate; stealing is when a whole box of the things disappear for re-sale at whatever we had before car boot sales were invented. The management can deal with the latter, but prefers to ignore the former. We are a happy, cheerful bunch of workers, and management wants it to stay that way.

Occasionally, I would get to cycle along the road to stores HQ – a couple of offices in a converted semi-detached house between the Victoria and Newark Works. Here work and social life was presided over by Cecil Marks, genial boss, neighbour in Lower Weston, and my best friend's dad (the same best friend who originally spotted the Stothert & Pitt link with Trader Jack's and who tried to teach me tennis at Newton Field). Surprisingly, this house still exists, I think, although it looks a lot cleaner. Or, if I was really lucky, I would be sent as far as the main offices at the Newark Works, where there were girls, a lot of them, all young, and some of them pretty too. I liked that, although when I got to work there for a few weeks, I missed the casual, male company of the Island, Bellott's Road and the Victoria Works. Yes, I got to work at the Victoria Works too, an invaluable insight into a heavy engineering firm at its peak. But most of all I remember the cycling, being between jobs, between bases, on my own, out of harm's way (except for cars, heavy lorries and foolhardy pedestrians). That stored up memory has perhaps stayed with me the longest and influenced both what I have done with my life since – and not done too.

I suppose we would have to categorise the Stothert & Pitt management in the 1960s as paternalistic rather than simply inefficient. Stothert & Pitt had been able to change back promptly in the years after the Second World War from military production to what they were good at – cranes, especially. In heavy industry, Britain led the world, and in cranes, Stothert & Pitt was simply that – cranes. A market leader with any incipient opposition trailing a long way behind. Or so they supposed. There was some diversification, especially into vibrating rollers and concrete mixers, but cranes still ruled our little kingdom.

Stothert & Pitt provided not just work, but also occupational health care, entertainments (a whole raft of sports and social clubs), a company magazine. Stothert & Pitt was a way of life for many of its employees. And many men died very quickly on leaving – a bizarre accident rate, you might say, but that was the truth of the matter. Others managed to pop off before retirement, an efficient way of avoiding the issue of what retirement might hold. Working at Stothert's gave shape and meaning to men's lives, perhaps more so than to most of its women employees, always more disposed to think of it as 'just a job', with the emotional focus of their lives elsewhere. Yet the emphasis on the provision of social facilities at Stothert & Pitt in the 1960s was not new. By the First World War, the Athletic

Founded in 1906, the Athletic Club was an important part of Stothert & Pitt. This Membership Card for the 1911/12 season belonged to Mr E. Dale. It is now in the extensive archive at the Museum of Bath at Work. The crane is still on the flag and badge of the Stothert & Pitt Rugby Club.

Club was well established, with grounds at Odd Down for football, rugby, cricket and quoits, and indoor facilities for boxing, billiards, bagatelle, cards, dominoes and draughts. Established in 1906, the club membership cards included the logo of the crane (the bird that is) still used by the Stothert & Pitt rugby club.

The work had its disciplines, of course. There was clocking in and out for a start. I still remember being docked half an hour's pay for a five minute late arrival. I only did it the once. Money was too valuable a commodity for a working-class student. At the end of the day, the shopfloor workers, who started before we did in the morning (although stores records were based in the factories, they were still 'office') would be cleaned up and ready well before 5.00 o'clock. They were allowed to clock off at five minutes to the hour, in order to be ready to leave at 5.00 on the dot. The resulting scrum as hundreds of men attempted to leave on foot and on bike at exactly the same moment is my abiding memory of the Victoria Works. It was industrial discipline of a sort, but enough to make any self-respecting Victorian industrialist turn in his grave.

The Stothert & Pitt sickroom, c.1950. Sister E.K. Botting (Victoria Works) takes notes. Dr E. Scott-White examines Jim Wellington's ear, watched by Nurse E. Cowley (Newark Foundry). Minor accidents were common, as in any heavy engineering works, but serious or fatal accidents very rare.

(left) Stothert & Pitt Rugby Football Club in 1903-4, the earliest known picture of a Stothert & Pitt sports team. From left to right, standing: C. Williams (treasurer), W. Comely, A. Fairweather, T. Gulford, F. Tipper (Somerset Rugby Union), J. Brunt, F. Cook, S. Golledge, J. Smith, W. Offer (secretary);
centre row, seated: A. Webber, T. Padfield, C. Spillane (captain), G. Padfield (vice-captain), A. Hodges;
front row, seated on the ground: J. Smith, H. Meddick, G. Jones.
The original photograph hangs in The Cranes' clubhouse at Corston.

(right) Major General Watkinson (S&P Managing Director) presents a football trophy (probably the Bath and District Knock-Out Cup) to Vic Anderson at Twerton Park (home of Bath City Football Club) in front of an extremely well-dressed crowd in the grandstand. The photo is a clear indication of how seriously the company management took such events.

Working at Stothert's

So what was life like at Stothert & Pitt during the second half of the twentieth century? From 1948 Stothert & Pitt started to issue a free monthly staff magazine to its employees. The date is significant. During wartime, Stothert's had been engaged mainly on war work. Ernest Bevin (an ex-docker from Bristol who must have been all too familiar with Stothert & Pitt cranes), Minister of Labour in the wartime government, ensured that there was industrial peace throughout the war, with worker rights respected and wage levels kept up in return for full co-operation with the war effort. The Labour government of 1945-50 (and many would argue the subsequent Conservative governments) attempted to keep this social and economic settlement going in the new conditions of peacetime. The NHS, the nationalisation of major industries such as the coal-mines and the railways, extensive slum clearance and house-building programmes were part of the new settlement, a social wage to complement actual wages from paid work. Volume 12 of *The Stothert & Pitt Magazine* was making more than just a local statement in 1959 when it claimed that:

> *Our aim is to encourage and engender a feeling of "togetherness" among the people who make up our organisation, and those who have joined the retired list. We want you, our reader, to be one of the family.*

Such paternalism was not new. In the nineteenth century, the Great Western Railway had combined stern industrial discipline with concern for the physical and moral wellbeing of its employees. The evidence is still there in stone, in the Swindon 'railway town' with its solid workers' houses, its church, its health clinic. Neither was that sense of 'family' an entirely spurious notion. Employment at Stothert's runs through the veins of so many old Bath families. In 1970 the staff magazine reported that Mr Simon Pitt, great-grandson of Robert Pitt, had been appointed as export manager in the Construction Equipment Division. A feature in 1966 celebrated 50 years of National Savings, and revealed the existence of a 'Stothert & Pitt Savings Group' with 802 members (about half the workforce). I certainly remember taking a few (old) pennies to junior school in the early 1950s when saving was a patriotic duty that linked family, school, employment and the state of the national economy. The contrast with today's 'credit' society, in which it seems that getting into debt is the badge of honour of every good consumer, could not be more stark.

The Stothert & Pitt Magazine specialised in up-beat good news stories, as company magazines have always done. But it went beyond this, with news of events around the city, poems, accounts of holidays abroad (a new phenomenon in the 1950s; these accounts are often highlighted by the sighting of Stothert & Pitt cranes on foreign docksides!), humour, technical information about new products and markets, and news of the growing range of sporting and social activities based within the firm. Raymond Leppard, then a rising talent in classical music, now a distinguished conductor, reported on the 1959 Bath Festival. The connection? Raymond Leppard, like Roger Bannister, had attended the City of Bath Boys School. I was there myself from 1956-63, and these two men were both held up as models of excellence we might follow. But the Stothert & Pitt connection? He was the son of 'our chief buyer, Mr A.V. Leppard.' Four years later Leppard senior retired. In his retirement speech, 'he concluded by saying he hoped he would be allowed to wander into the Offices and Works once in a while just to keep in touch with old friends.' I remember such visits taking place during my time at Stothert & Pitt; old colleagues, however busy, could always find a few

S&P Magazine, Autumn-Winter 1959. The cover illustration is in an aggressively contemporary style which matches the company's push for increased exports and new markets for its products.

S&P Magazine Summer-Autumn 1964. A timeless scene on the Kennet & Avon Canal at Widcombe, with an early Stothert & Pitt iron footbridge in the background.

minutes to stand and chat. The 'family feeling' was not just an invention of management.

Labour relations were a key issue within business management, just as they are today. There are only occasional glimpses of this in the staff magazine. In general Stothert's management seem to have been very successful in maintaining good relations, though some might want to question the cost of the generous social and leisure facilities and the sometimes rather relaxed approach to work. In 1959 the centenary dinner of the AEU (Amalgamated Engineering Union) was held 'in our New Canteen' (what pride those capital letters contain). Let us see what was said:

> Mr F. Tansey, President of the Bath No 1 Branch of the Union, in proposing the toast to the guests, mentioned the close and happy association that had always existed between the Union and Stothert & Pitt Ltd.

Certainly, if we compare Stothert's with the British car industry of the period, with its unofficial shop stewards' movement and its lightning strikes, as reflected in the 1959 Peter Sellers film *I'm All Right Jack*, there was a great deal of harmony. Five years later when Mr Tansey retired from the Machine Shop, he was described as AEU Senior Shop Steward and a governor of Ruskin College, Oxford. Ruskin, with its generous scholarships for working-class students, had educated several generations of trade union leaders, as well as creating a ladder for those who wanted to move on to professions such as teaching, or into left-of-centre politics.

Management had thought deeply about these matters. In 1963 the magazine quoted an extract from an influential book of the period, Michael Shanks' *The Stagnant Society*:

> The problem is not one of economic management but of social engineering; we need to adjust the social framework so as to overcome the social, psychological, and political frictions which are at present stopping the economic techniques from working effectively.

It makes curious reading to a modern reader, a quarter of a century after Margaret Thatcher made up her mind that what British industry needed was not 'social engineering' but the political will to reconstitute the economy in a new way. She unleashed a whirlwind which blew away many cherished British institutions, including Stothert & Pitt. But at least there was acknowledgement in 1963 that a problem existed. That was the nub of the contradiction in the paternalist firm with its model of the one big happy family: just as in a family, the interests of parents and children are locked together but very different, so in a firm the concerns of management have seldom matched the day-to-day concerns of the majority of the workforce.

The still distant economic crisis of late twentieth-century Britain was referred to rather obliquely in the staff magazine. Productivity and exports were major concerns. In 1960 there was an editorial on the reduction of working hours from 44 to 42 hours per week, warning that 'to hold costs at their present level, an

Presentation of Amalgamated Union of Engineering Workers long-service medals to Harold Webb and Ron Bishop, Stothert & Pitt AUEW representatives. On the left is George Sanderson, District Secretary; centre, the Divisional Organiser; and right, John Woodward (Tool Room shop steward and District President). (Bath Chronicle)

increase of 7% in output is needed from everyone involved.' Keeping costs stable was crucial in the face of rising international competition, as other countries recovered from the war and sought to match, if not surpass, their British competitors. There were increasingly frequent features in *The Stothert & Pitt Magazine* about the export drive and new markets opening beyond the fast disappearing British Empire – in South America, the Gulf and Europe. Of course, the firm did have some very specific problems, not least that many of the premises laid out along the Lower Bristol Road were liable to flooding. This problem reached its peak in December 1960 when the Avon rose 20 feet above its normal level. If Stothert & Pitt records are hard to find, it is as much to do with a century of flooding as the cunning with which Robert Maxwell covered his tracks in the 1980s.

One noticeable feature of the staff magazine is a genre that we might term 'exemplary lives' – accounts of the lives of employees, usually at or about retirement, both in and out of the firm. One such is Mr W.E. (Bill) Shepherd, captured in 1959 on his retirement at the age of 70. Bill Shepherd had worked for Stothert's for half a century. He had also been a leading member of the Salvation Army in Bath, which he joined in 1905. In addition, he had been active in the St John Ambulance Brigade, the Royal United Hospital (in a number of voluntary capacities) and the Stothert & Pitt Benevolent Fund. As we read:

> *Attendance at meetings of these bodies required a number of leaves of absence, and he speaks warmly of the full co-operation that the firm gave to enable him to carry out his duties.*

Good worker, upstanding citizen, we have the model worker of the middle of the last century. It is a world that seems a long way away from us now.

Presentation by George Webber (foreman) on the retirement of Vic Walterson from the Machine Shop in 1967. On the left of George Webber are Jack Mogg, Mr Jefferies (first name not known) and Ken Pearson. To his right are Ron Foulkes, Stan Eyles, Michael Fear, Harold Plumley, two unidentified men, then Ernie Dyrall and Jackie Thomas. Ted May is on the extreme right of the picture. Others in the picture include Bernard Jefferies and George Moger.

It was not only the economy that was changing but society too. Reports of annual beauty contests (presumably female beauty only) gradually gave way to reports of women being taken on in professional jobs, even the odd apprentice or two. And then a six-a-side football team to challenge the lads in 1975.

In 1967 a retirement photograph taken at Weston Island shows two black employees, reflecting the increasingly multi-ethnic nature of the city. A few years before, in 1963, the West Indian cricketer Gary Sobers had scored a brilliant century, playing for West Indies against Somerset on the Rec – the same Recreation Ground where Bath Rugby play in the winter. I was there, at least for part of his innings, having rushed straight down from Beechen Cliff after school. Such was the way in which we prepared for 'A' level exams in those days. So too was a noisily ecstatic bunch of Bath West Indians. In 1976 there is feature about a Jamaican secretary with a university degree in French and Italian. There is some indication in the same copy of the staff magazine of the wartime origins of the British West Indian community – a Mr St John Dick who came to Britain in 1943, and had played cricket for Stothert & Pitt. In 1976 he was running a Bath Multi-Racial Club with 120 members, part social club and part advice centre. But there were no black faces in that year's intake of apprentices.

The impact of foreign competition is a constant in *The Stothert & Pitt Magazine*. In Summer 1963 there was a report about 'the smooth, safe SP33 – a crane that gives Britain the lead'. In fact it clearly didn't give Britain the lead, since the article goes on to bemoan the previous lack of competition for comparable French and German models, and the fact that

> Had the company taken this step five or six years earlier, when they were I developing their highly successful dockyard crane, there can be no doubt that the reception would have been overwhelming.

At that year's AGM it was reported that

> overseas markets traditionally supplied from Britain were increasingly subjected to attacks from other countries.

More and more, the existence of Stothert & Pitt as an independent company was being thrown into doubt. This surfaced in humorous form in an article on takeovers in 1969. The author suggests the Bayer factory, the bricks of which nestle against the stone of the Newark Works, as a possible takeover 'partner', on the basis that both products (corsets and cranes) are 'uplifting'. Is the joke so awful that some kind of 'insider' knowledge is suggested? By 1971 the Chairman was denying in the staff magazine a *Guardian* newspaper story that a merchant bank (Rea Bothers) was buying a stake in the company with a view to future asset-stripping. These were crisis years at Stothert's. In 1972 it was reported that 'after five years struggling to make a profit and remain solvent, Stothert & Pitt has turned the corner at last'. Precisely what corner had been turned is not explained. The merchant bank story resurfaced in 1975 with the announcement of a new Chairman, Mr Sam Wainwright, Managing Director of Rea Brothers. It was innocently added that he had been on the Board since 1970!

Despite all this, there is an underlying optimism. In 1968 the company opened the Avalon Training School at Twerton, and continued to take on good numbers of apprentices. In 1976, the intake was upped to 30, at a time when 'other firms' were cutting back their intakes. I know they were. At the time, I was a young lecturer in a technical college in a prosperous part of the country. The college was threatened with closure because of declining numbers of apprentices to fill our day- and block-release courses. While I am not suggesting that Stothert's fiddled while the engineering industry burned, there is a certain irony in the detailed coverage in The

Stothert & Pitt opened an Apprentice Training School at Twerton, near the Weston Island site, in 1960. Beside training on the job, apprentices were expected to study for engineering qualifications at college. In this photo, Mr A.H. Yates, the Principal of Bath Technical College, is presenting an award to apprentice Mr T.C. Chivers on 6 November 1960. Even in the 1980s, Stothert & Pitt was still taking on apprentices.

Stothert & Pitt Magazine of the problems of the Sports and Social Club. In 1970 they attempted to increase the weekly subscription from two old pennies to six. The magazine reported that only 400 out of 1,100 members had signed the necessary authorisation for deduction from salary, and that some older members still resented the 1940s increase from 1d to 2d! By the following year a recruiting drive was going well, based on two new pennies per week, a full 20% less than the sum demanded in the old currency. A smart new bar and club-room was being inaugurated at the Newton Field. The employees of Stothert & Pitt were walking into history, their eyes firmly shut.

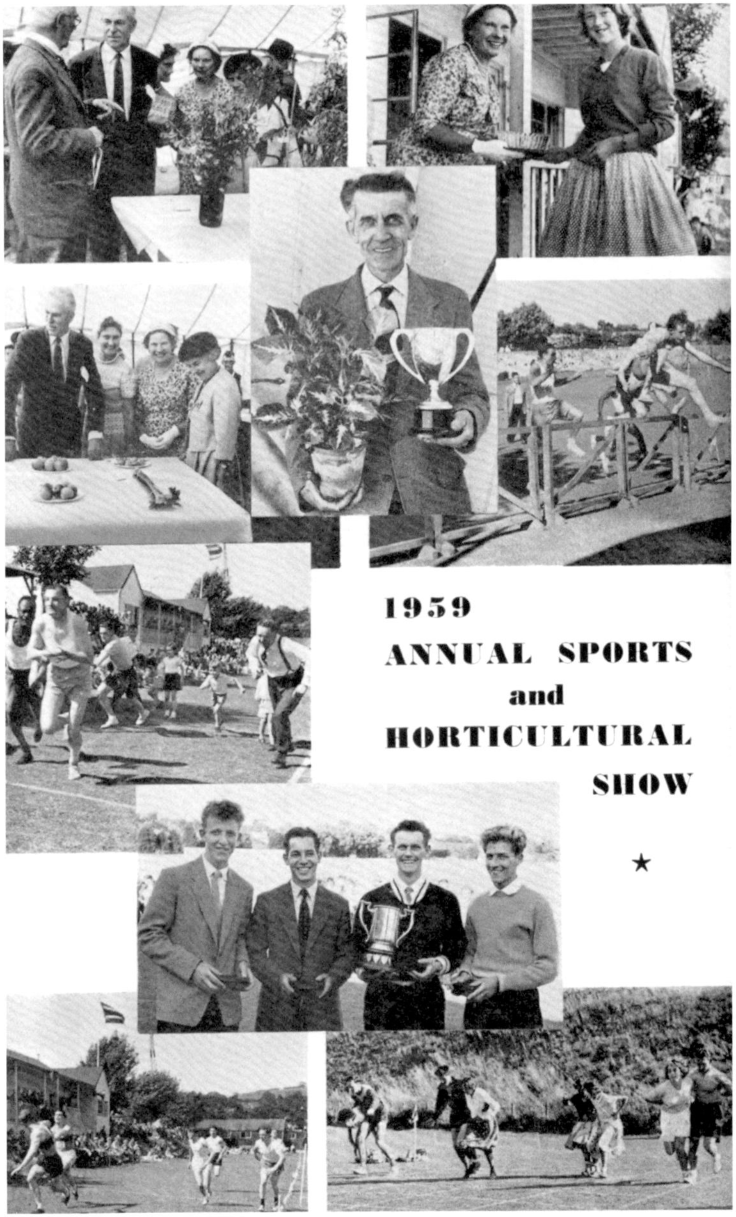

1959 ANNUAL SPORTS and HORTICULTURAL SHOW

★

A compilation of photos of the important annual Sports and Horticultural Show from The Stothert & Pitt Magazine. The building is the cricket pavilion at Newton Field, still standing.

32

(right) Mullah's 8 team, winners of the Stothert & Pitt skittles league, July 1960. On the left is Charlie Brooks, who grew chrysanthemums and won many prizes at Stothert & Pitt shows. Then Sam Salter, Bill Wynes and Tom Hooper. Holding the shield is Ron Lewis, nicknamed the 'Mad Mullah' for his antics, who gave the team its name. Behind him are 'Smudge' Smith and Eric Wynes, then Phil Richards, Wally Cox, Brian Wynes (Bill's son) and Dave Craddick. The Skittles League was a central part of the social life of the firm. This team played at the Belvoir Castle pub.

(below left) Another shield for the Mullah's 8 skittles team showing many of the same players. The boy is William Hooper, son of Tom Hooper (second from left). (William Hooper collection)

(below right) Stothert & Pitt Bowling Club skittles team at Twerton Liberal Club, Brougham Hayes. From left to right, John Skinner, unidentified, John Hosey, Bryan Gould, Len Hughes (captain) holding the trophy, unidentified, Cecil Marks, unidentified. John Skinner's brother, Ken, managed the Avalon Apprentices School. Cecil ran the Stothert & Pitt skittles league. (Bath Chronicle)

Retirement presentations

(below) A large gathering of workers to mark a retirement presentation to Edie Bishop by George Webber in 1967. The building is the Pump Shop of the Victoria Works, facing on to what used to be Longmead Street. The three men on the left are Ivor Weight, Joe Rickards and Sam Salter, with Les Smith in the white coat. From the far right we have identified Ted Nixey, George Moger, Jackie Thomas, Tommy Twine, Ken Pearson and Harold Atkins. On Harold's left are Maureen Wyatt and Jeannie Messer. Other workers in the photo include Les Norris, Bert Harding, Jimmy Tucker, Ken Twiss, Colin Latham, Derek Martin, Sandra Edgell and Linda Pegler (immediately behind Edie Bishop), Bob Perkins, Len Walters, Colin Bletso, Pat Messer and Roger Ferris.

(right, above) Retirement presentation to Reg Ward of the Pump Fitting Shop by Jim Murray, 1971. The three figures on the left are Ron Tipper, Jack Kimber and Ray Stacey. Ted Nixey is next to them with his hands clasped. Others identified in this picture are Pete Prewitt, Les Trubody, Dave Eddy (crouching in bowler hat), Pete Couzens, Ray Lea (in jacket and tie), Cecil Grubb, Bob Webb and Len Walker (extreme right). (Ted Nixey collection)

(right, below) This photograph was taken in the Victoria Works yard on Tom Hooper's retirement day in 1978. He drove one of the firm's Lister trucks. Across the river is the old TA Centre, now replaced by Victoria Bridge Court. (William Hooper collection)

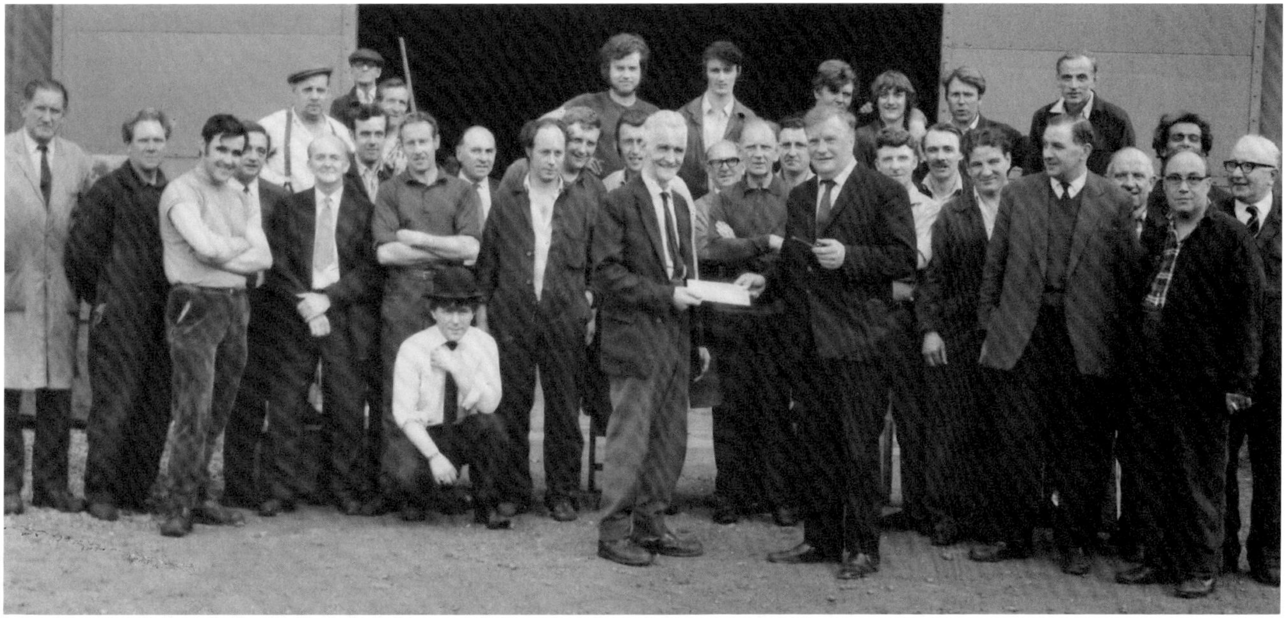

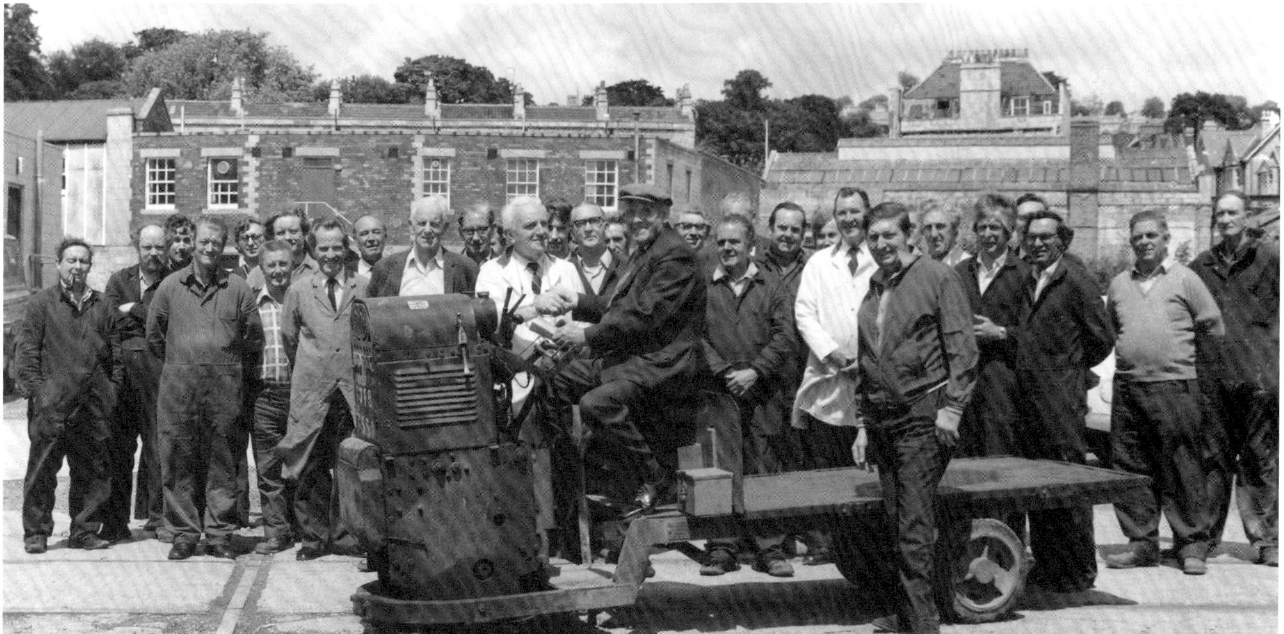

A Tool Room presentation by Reg Clark, charge-hand, to Ellis Kite, foreman. Front row, from left to right: Geoff Dix, Sam Wood, Freddie Jones, unidentified S&P pensioner in hat, Norman Eames, Bill Day, Brian Higgs, Albert Goodfield. Second row, from left to right: Frank Gulliford, George Hedges, Fred Bradfield, John Woodward, Vic Cook (charge-hand, in white coat), Paul Willmott, George Jewiss (in tie), Bill Densley. Heads peeping through behind this row are Ron Maslen, Vic Bacon, Roger Evans, Sam Warren, Bill Roberts, Harry Evans, Bob Styles and one unidentified man. Back row, from left to right: Len Luff, Laurie Chiswell and Nick Grenfell, with Mike Rogers balanced right at the top. Many Tool Room workers wore collars and ties under their brown smocks as a mark of their status in the firm. (Brian Higgs collection)

Retirement presentation to one of Stothert's most popular characters, George Wakefield, who went to live in Falmouth. This is a detail of a much larger photo containing over 50 people. On the left are Danny Owen and Paul Derrick. The man in the suit, half hidden behind the man in the white coat, is Charles Norwood. Ray Butt can be seen over his left shoulder. On the far right are Nigel and Dave Barnes. (Dave Barnes collection)

Presentation by Ken Gray (foreman) on the retirement of Jim Gregory, Structural Shop, 1967. From left to right are Adrian Beck, John Jefferies, unidentified, John Dyke, Bill Vines, Johnny Glisson, Jim Gregory, Ray Butcher, Bernard Webb, Ken Gray. The next three have not been identified. The tall, bearded man is Johnny Willcox who now lives in Oregon, and the man on the extreme right is Pete Ball.

Leaving presentation to Maurice Cottell by George Watson (Head of Progress Department). From left to right are Don Withers, John Skinner, Vic Chapman, Rosemary Dancy, Keith Letts, Maurice Cottell, Maurice Counsell, George Watson, Doug Lay, unidentified, Cecil Marks, Janet Oram (née Dolman) and Sid Blackmore. Maurice Cottell played double bass for a number of local bands, including Acker Bilk's Paramount Jazz Band. He learned the instrument during the war in Colombo, where double bassists were in short supply. Later he joined up with Joe Brickell to form the Spa City Stompers and then the Joe Brickell Jazzmen. Later still, he played with George Watson on drums and Don Withers on piano, playing for dances at many venues including the Pump Room. John Skinner often travelled with Maurice (or Mark as he was called in the music world) and remembers the tight squeeze of two men and a double bass in an Austin Seven.

This view of the Preparation Bay of the Structural Department is one of a number of photos taken inside the works during the last few days before redundancy in 1989. Second from the left is Josh White, the works convenor for the Electrical Trade Union (ETU). Fourth from the left is Fergie Smith, then Joe Lock and Keith Abrahams. The foreman in the white coat and the two men either side of him remain unidentified. The next man is Ivor Gibbs, then Dave Barnes (with hands clasped), then Johnny Adams and Denis Gooding. (Dave Barnes collection)

WEDDING PRESENTATIONS

(below) Wedding presentation to Tony Mitchard, No 2 Machine Shop, by Norman Fisher. From left to right are John Gould, John Harris, Wally Lewis, Eric Fisher, Graham Button, Norman Fisher, Lesley Evans. The two men at the back behind Tony Mitchard have not been identified. To Tony's right are David Humphries (nicknamed Frankie Vaughan for his hair-style), while Tony has his arm round popular Welshman Ivor Weeks. The man in the leather jacket is Michael Fear and on the extreme right is Pete Edgell. Right at the back behind Michael Fear is Barry Tovell, but the other two men on the right are unidentified.

(right, below) Presentation to Wally Cox, by Jack Mead (foreman) in 1949. Pictures of this type were to appear in every S&P Magazine and represent the cross-over point between work and family life. A very young Dave Barnes is on Jack Mead's left, with Reg Fellowes to his left again. The two men in hats are Bert Rummings (head foreman) and Albert Webber. (Dave Barnes collection)

(right) Wedding presentation to Pat Carless (née Jones), a bonus clerk in the Structural Shop, 1960. Making the presentation is Mervyn Humphries (rate-fixer). From left to right are Dave Barnes, Harry Newman, Eileen Ball (née Tippett), Joan Curtis (née Hart) Reg Hallett, Molly Snowden, Pat Brown. The next woman we only know as Dot, while between her and Mervyn Humphries is Ann Jones (Pat's sister). To the right of Mervyn Humphries are Carol Ashton, Pat Carless, two unidentified women, Fred Cavill and Bill Vines.

(above) The Spectres played at the annual S&P children's party held in the Canteen. This was called the 'Firemen's Christmas Party', as it was organised by the firm's internal fire brigade. Those attending included employees' children and children from local children's homes. (Brian Higgs collection)

Eddie and Ruth Barlow at a Gardening Section social at Smith's Assembly Rooms in Westgate Buildings, 1960. The Barlows and their five children later emigrated to Australia. Sitting on the left of the photo is Alf Smith, while to the right of Mrs Barlow is Colin Vuagniaux, who is remembered as always retaining his youthful look.

A Labour of Love? Careers at Stothert's

At the same time as the announcement was being made of the end of Stothert's as a manufacturing firm in 1989, an oral history project started at the Museum of Bath at Work. Close on 200 audio recordings were made of Bath people talking about their working lives. Given the importance of Stothert & Pitt as a local employer, it is not surprising that Stothert's is well represented in the archive. For some, Stothert's was their whole working life, for some just one part. The Second World War made an important incursion into careers, as did other national happenings, especially the rise and fall of the economic cycle, slump and boom. These recordings emphasise, above all, Bath as a working place, a work where the drama of ordinary lives was lived out much as it was in other manufacturing towns around the country. There is little or nothing here that might make one think 'World Heritage Site'.

Mr Smith worked for Stothert's from 1963. He first visited the firm as an employee of an electrical firm to install cables in the boiler shop. He has left a dramatic account of his first impressions of a large engineering works:

It was in the boiler shop, and it was really filthy and frightening, frightened me to death, cranes going overhead, the travelling cranes, carrying heavy objects, never seen 'em before, it was really frightening.

Once he had joined the firm, he took to it:

It was a good atmosphere, it was really good. The wages was low, poor wages ... Apart from the wages, one thing about going into Stothert's, you had a job for life. You weren't going to get the sack, you had a job there, it was security.

Security, a good spirit about the place. These were the things employees valued. But Mr Smith rejected the impression given by *The Stothert & Pitt Magazine* of 'one big family', one big army of workers, the impression given by the Canteen and the Sports and Social Club. Most employees kept to their own corner of the firm:

I think that's a false impression, really, like I worked in the Boiler Shop, and I wouldn't know hardly anybody that was working across the road in the Machine Shop.

As for many others, employees, friends, ordinary residents of the city, the 1980s was a puzzling time:

They did get a bit low on orders in the latter stages, but I don't know why it folded up, or why Maxwell bought it. I mean, you hear a lot of different stories, why the firm was going under. The funny thing was, after they finished, within a month, talking to one of the draughtsmen who still works in Stothert's ... I was told that a fortnight after they had orders coming in for cranes. It's a very funny business, I think it was just people high up making money.

(By 'still works in Stothert's' he was referring to the small design and spares firm that emerged from the rubble of Stothert & Pitt.) He was philosophical about what happened, but perhaps at his age could afford to be:

'Did people start panicking?'
'Some did. Personally, I didn't. Actually, previous to that I'd had a bit of heart trouble, 1981. Lifting, I strained the heart, and after that I decided I was going to take things easy, and I did. So to me, when it came along, it wasn't a great shock. I said, "Right, I've done 45 years' work, now I'm gonna have a rest".'
'And was it a great shock?'
'It was to a lot of people.'
'Nobody expected it?'
'No ... but it's the way people take it, like I was resigned to the fact, the last few years I wasn't gonna work hard. When I finished, I was going to be able

to walk out. Whereas I knowed a neighbour of mine, lived along Twerton, near me, had to retire ... and the week he retired, he dropped dead with a heart attack, because he was so used to work, continually going to work, it occupied his mind all the time. He was so worried about what he was going to do when he retired that he didn't get the chance.'

It must have felt very different, too, for the couple of dozen apprentices shown the door inside the first year of their apprenticeships.

If life at Stothert's post-Second World War was quite relaxed ('a lot of people had very soft jobs', was Ted Nixey's verdict on 50 years at Stothert's), others remembered a strong tradition of work discipline. Fred Harper, who began at Stothert & Pitt as an apprentice in 1937, remembered the lack of either tea-breaks or a canteen. He used to bike home to Larkhall for his dinner, 15 minutes each way out of a dinner-break of one hour. Clocking-in had not yet arrived, instead:

Each worker had a numbered disk which had to be taken off a board and dropped into a box. Hooters went at 7.30am and at two minutes to 8.00 o'clock. Even if you were only a minute late, you lost half-an-hour.

On Friday evenings they were paid in works number order – if you were late you had to wait for your number to come round again. Fred was a reluctant Stothert & Pitt man. He had wanted to join the Navy but his Dad had insisted he learned a trade first. The war gave him his chance. From Dad's Army – the Home Guard – he moved into the Navy, and was able to complete his apprenticeship. 'I learned more about engineering in the Navy,' he claimed, 'than in Stothert & Pitt.'

Moving further back in time, Mr Coles started work at Stothert & Pitt in 1920. His father also worked there, although even further back both sides of the family had been weavers. His step-grandfather was at Carr's Mill in Twerton and his grandmother came from a weaving family at Wellington in West Somerset. Mr Coles followed a model apprenticeship through the Newark Works, the Victoria Works and into the Drawing Office, presided over by the genial Colonel Pitt. Unlike other Stothert's men, Mr Coles moved on, eventually becoming the right-hand man to Mr Cross, the inventor and manufacturer at Midford Road, with his rotary valves, his steel piston rings (so in demand in the Second World War that aircraft factories sent cars down each week to collect them), his 'humane sheep-killers' and his wax milk containers. Cross Manufacturing is still very much in existence and Mr Coles was instrumental in the formation and running of the company. The firm is in Midford Road opposite what was St Martin's Hospital and is a very successful engineering firm, a comparative rarity in Bath today.

The industrial heritage of Stothert & Pitt

How do we measure 'industrial heritage'? Sometimes it may be specific remnants of an industry; at other times a memory in the popular imagination. Such memories are not trivial. Public memory helps to define for people both who they are and where they are in a changing and uncertain world. Only a few miles south of Bath, largely unknown to most visitors to Bath and even to many of its more recent residents, lie the remnants of a substantial coalfield where coal was still mined until the 1970s. Only the batches (slag-heaps) in places such as Radstock and Pensford and the 'volcano' near Paulton (a conical-shaped batch which still dominates that area) bear silent witness to a large-scale industry. Yet that is only partly true, as coal-mining has also contributed a major part of one of the best industrial museums in England – Radstock Museum in its finely restored 1898 market building. People are proud of their museum, for the long history of coal-mining in North Somerset remains an important part of how local people still identify themselves. Like the use of spa waters in Bath, the use of coal goes back to Roman times.

Some of the Stothert & Pitt buildings still stand: the Oak Street office-block is used by Beazer Homes; Weston Island is now a garage for the local bus company. Only one site is still in use for engineering – Bellott's Road, where Gerry Matthews, himself an ex-Stothert & Pitt apprentice, runs M&B Engineering: 'fabrications and welding in mild and stainless steel.' The largest site of all is the Victoria Works, now designated as the Western Riverside and subject to a large-scale redevelopment scheme which will include the gas-works and the gas-holders, for so long dominant features of the city (and also of my Lower Weston childhood). This redevelopment is almost entirely about housing, give or take a few shops. House prices in Bath are probably the highest in the country outside of London, and the logic of the property market excludes all other possibilities. More imaginative possibilities, such as using one of the old gas-holders as the framework for an international concert-hall to match the pretensions of the Bath International Music Festival, have not even reached the starting-blocks, while planning approvals are being actively sought for the largest housing development in Bath since the postwar council housing boom.

The main focus of debate about Stothert & Pitt and living memory centres on the Newark Works. But in order to understand the controversy around this building, we need to look briefly at the history of another West Country engineering firm – James Dyson at Malmesbury

The Bellott's Road site of S&P, now M&B Engineering, 2006. Gerry Matthews (inset), one of the partners, had served his apprenticeship at Stothert & Pitt. It is the only Stothert's site still being used for engineering production. (Author's collection)

in Wiltshire. Increasingly frustrated by the relatively high level of wages in the UK, and the problem of finding suitable suppliers of parts as other engineering firms closed, Dyson decided in 2002 to move manufacture of vacuum cleaners from Wiltshire to Malaysia. In common sense terms it makes very little sense to manufacture products so far from the places where they will be bought and used. In economic terms it makes every sense, and the increasing level of Dyson profits since is evidence for this. Dyson himself, knighted in January 2007, claims that the firm now employs almost as many people as before in the company headquarters and Research and Development Centre at Malmesbury. But of course they are not the same people. Semi-skilled workers on an assembly line do not move into office or research jobs.

Some 590 jobs were lost in 2002 and a further 50 in 2004 when washing machine manufacture was also moved out of Malmesbury. One of the Dyson employees, Rob Lewis, was reported in the local press making similar statements to those made by Stothert & Pitt employees:

> I feel like I've been on death row for the last seven months. Tonight, we're going in for the actual execution. Everybody's just sad, everybody's down.

There is no doubt that such a closure in a small Wiltshire market town would have had an even more devastating impact than a closure in the much larger city of Bath.

Dyson insists that his company is still a British company and he remains deeply involved in the continuing debate about the future of productive industry in the UK. He resigned as Chairman of the Design Museum in 2004, insisting that 'we have no choice but to shake off our obsession with styling'. He sees engineering at bottom as the production of useful objects that people want to buy and use. Dyson aims to breathe new life into engineering by setting up the Dyson School of Design Innovation in a six-storey building designed by Chris Wilkinson, twice winner of the Stirling prize for architecture. And where will this school be? Well, somewhere called the South Quays in Bath. It took some time for the penny to fall that 'South Quays' meant the Newark Works of Stothert & Pitt (crane-makers to the world) in the Lower Bristol Road!

Dyson will certainly contribute to the cost of the project, though there will also be a substantial public contribution through the Regional Development Agency and central government. Likewise, running costs will be covered by the Learning and Skills Council. It is disappointing that reaction within the city has concentrated on the merits and demerits of Wilkinson's designs, on the one hand, and the newly discovered architectural merits of the Newark Works on the other hand. The Newark Works, occupying a narrow and difficult site between the River Avon and the Lower Bristol Road, is an early work by Thomas Fuller, who went on to build bigger and perhaps better buildings in Canada. In an apparent compromise move in late 2006, new plans were submitted which showed the retention of the façade of the Newark Works and the new building fronting the river and city. Soon afterwards, English Heritage listed the building, in response to lobbying from campaigners. Yet at no point does there seem to have been serious consideration of including in any plan for this site, or for that matter the Victoria Works/Western Riverside site, any substantial reminder to the people of Bath about what Stothert & Pitt actually built – cranes. To find Stothert & Pitt cranes it is necessary to go some miles downstream to Bristol. Outside the Bristol Industrial Museum (scheduled to re-open in 2009 as the Bristol Museum) are four splendid rail-mounted cranes, labelled '1951, Stothert & Pitt, Bath'. A little further along towards the *SS Great Britain* is an 1878 Stothert & Pitt 'Fairbairn' crane, its jib framing a splendid view of Bristol Cathedral and Cabot Tower. It is in working order too.

The steam-driven 'Fairbairn' crane.

(right) The crane is illustrated in the 1885 catalogue and is shown in situ on the Floating Harbour in Bristol. (British Library, London)

(below) Restored by Bristol Industrial Museum, the actual crane can be seen in its original location today. The view in this photograph looks across the harbour to Bristol Cathedral and the modern commercial buildings which are part of Bristol's revival. (Author's collection)

STOTHERT & PITT – REFUSING TO DIE

Yet despite everything, Stothert & Pitt does still exist. The nearest landmark is the *Belvoir Castle*, where my Twerton uncles used to drink, and behind which Bath City Football Club used to play before their ground was taken over, soon after the end of the First World War, by Stothert & Pitt. Lurking behind the row of garages on the Lower Bristol Road between Midland Road and the Windsor Bridge Road, backing on to the wilderness of the gas-works site, are the current offices of Stothert & Pitt. I would emphasise offices, for nothing is actually made here.

The firm did not disappear entirely in 1989. It is as if in hiding. Here a couple of dozen employees are using both the original drawings for Stothert & Pitt products and their own design skills to respond to the handling needs of clients. No vibrating rollers or concrete mixers; Stothert & Pitt has returned to its core business of cranes. The firm belongs to a much larger crane group – Clarke Chapman, based in Gateshead – which in turn is owned by a firm called Langley Holdings. This is described on its website as a 'global multi-disciplined engineering concern serving a multitude of world-wide markets, employing many hundreds of talented and committed people across the globe.' The mystery, as with most holding companies, is who does the making and where do they do it.

The continuing existence of Stothert & Pitt in this particular form, like the continuing and rather more profitable firm of James Dyson, marks a change from integrated ownership, design, manufacture and administration to one in which functions can be endlessly farmed out through interlocking subsidiaries, contractors and consultants. It is a hollowed out world in which the central fact of engineering – the transformation of raw materials into useful objects – seems suddenly peripheral to the operation. The cranes may still have 'Stothert & Pitt Bath England' on them; some of the employees may be survivors from old days; but is it really the same firm? I think not.

And as a footnote to this small contribution to the history of Stothert & Pitt of Bath, there are also a number of survivors from the social side of the firm. The Camera Club meets at the St John Ambulance H.Q. in Pulteney Mews. The rather higher profile Stothert & Pitt Rugby Club ('The Cranes') has its smart new ground and club-house, financed by the members, between Newton St Loe and Saltford. The club celebrated its centenary in 2004. The crane on its logo may have changed its colour from red to yellow, but is the same bird that adorned Mr Dale's 1911/12 Stothert & Pitt Athletic Club membership card, now in the safe hands of the Museum of Bath at Work.

At Newton Field, which the Western Riverside planning enquiry of 2006 decided should remain within the green belt, the Stothert & Pitt cricket club still plays, and leases out football pitches to a number of local clubs. The Stothert & Pitt Bowls and Tennis Club thrives and grows. The Sports Lottery Fund has paid for a new bowling green and for the conversion of the old caretaker's bungalow into a clubhouse. Sadly, only five of the members are now ex-employees. One of them is Dave Barnes who worked at Stothert & Pitt for 40 years, ending up as a trade union Works Convenor. He still lives in East Twerton, supports Bath City Football Club and takes a lively interest in the doings of the hundreds of other Stothert & Pitt employees who continue to live in and around Bath. It is a network of men and women that is quite invisible to the many thousands of new Bath residents who know nothing of Stothert & Pitt, but which continues to provide friendship, practical support and a sense of belonging in the city.